美国国家地理
双语阅读

Coral Reefs

珊瑚礁

懿海文化 编著
马鸣 译

第三级

外语教学与研究出版社
FOREIGN LANGUAGE TEACHING AND RESEARCH PRESS
北京 BEIJING

京权图字：01-2021-5130

图书在版编目（CIP）数据

珊瑚礁 ：英文、汉文 / 懿海文化编著 ；马鸣译. -- 北京 ：外语教学与研究出版社，2021.11（2023.8 重印）
（美国国家地理双语阅读. 第三级）
书名原文 ：Coral Reefs
ISBN 978-7-5213-3147-9

Ⅰ. ①珊… Ⅱ. ①懿… ②马… Ⅲ. ①珊瑚礁－少儿读物－英、汉 Ⅳ. ①P737.2-49

中国版本图书馆 CIP 数据核字（2021）第 228173 号

出 版 人　王　芳
策划编辑　许海峰　刘秀玲　姚　璐
责任编辑　姚　璐
责任校对　华　蕾
装帧设计　许　岚
出版发行　外语教学与研究出版社
社　　址　北京市西三环北路 19 号（100089）
网　　址　https://www.fltrp.com
印　　刷　天津海顺印业包装有限公司
开　　本　650×980　1/16
印　　张　37.5
版　　次　2022 年 3 月第 1 版　2023 年 8 月第 4 次印刷
书　　号　ISBN 978-7-5213-3147-9
定　　价　188.00 元（全 15 册）

如有图书采购需求，图书内容或印刷装订等问题，侵权、盗版书籍等线索，请拨打以下电话或关注官方服务号：
客服电话：400 898 7008
官方服务号：微信搜索并关注公众号“外研社官方服务号”
外研社购书网址：https://fltrp.tmall.com

物料号：331470001

Table of Contents

City Under the Sea

The shallow ocean waters look calm. But under the surface, a coral reef is a busy place.

Q What did the coral reef say to the water?

You're so shallow. 

Hundreds, even thousands, of different creatures swim and hide along the reef. There they find food and shelter. They make their home in this "city under the sea."

A coral reef is a very important ecosystem (EE-koh-sis-tum). More sea creatures live along coral reefs than in any other part of the ocean.

Reefs are found in many spots around the world. Most reefs grow in shallow, clean ocean waters on either side of the Equator (i-KWAY-tur). They need sunlight and warm temperatures year-round to survive.

Reef Talk

ECOSYSTEM: All the living and nonliving things in an area

EQUATOR: An imaginary line around Earth halfway between the North and South Poles

Reef Builders

Coral reefs
look like they are
made of rocks. But, in fact, they are
groups of animals called corals.
Each coral group is made up of many
separate coral polyps (POL-ips).

In reef ecosystems, there are two kinds of coral: hard and soft. Only hard coral polyps form reefs. They're named for the hard skeletons they build at the base of their soft bodies.

Hard corals make reefs.

Soft corals do not make reefs.

Reef Talk

CORAL POLYP: A small, simple sea animal with a tube-shaped body and a mouth ringed with tentacles at the top

One coral polyp can be as small as the head of a pin. But when many polyps join together, they make a reef that can stretch for miles.

Fan coral with open polyps

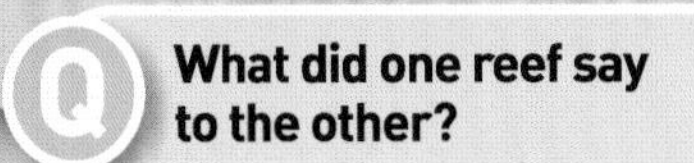

What did one reef say to the other?

Don't be so hard on yourself.

New polyps build their hard skeletons on top of old ones. Over many years, these layers of skeletons slowly grow into a coral reef.

Hard coral reef

Neighbors on the Reef

Creatures big and small can be found on reefs around the world. Sea stars travel slowly along a reef's surface. Giant clams rest there, too. Tube sponges stretch up from a reef like small chimneys. Seahorses wrap their tails around pieces of coral. Sea turtles swim around reefs. Eels hide in a reef's cracks.

Sea star

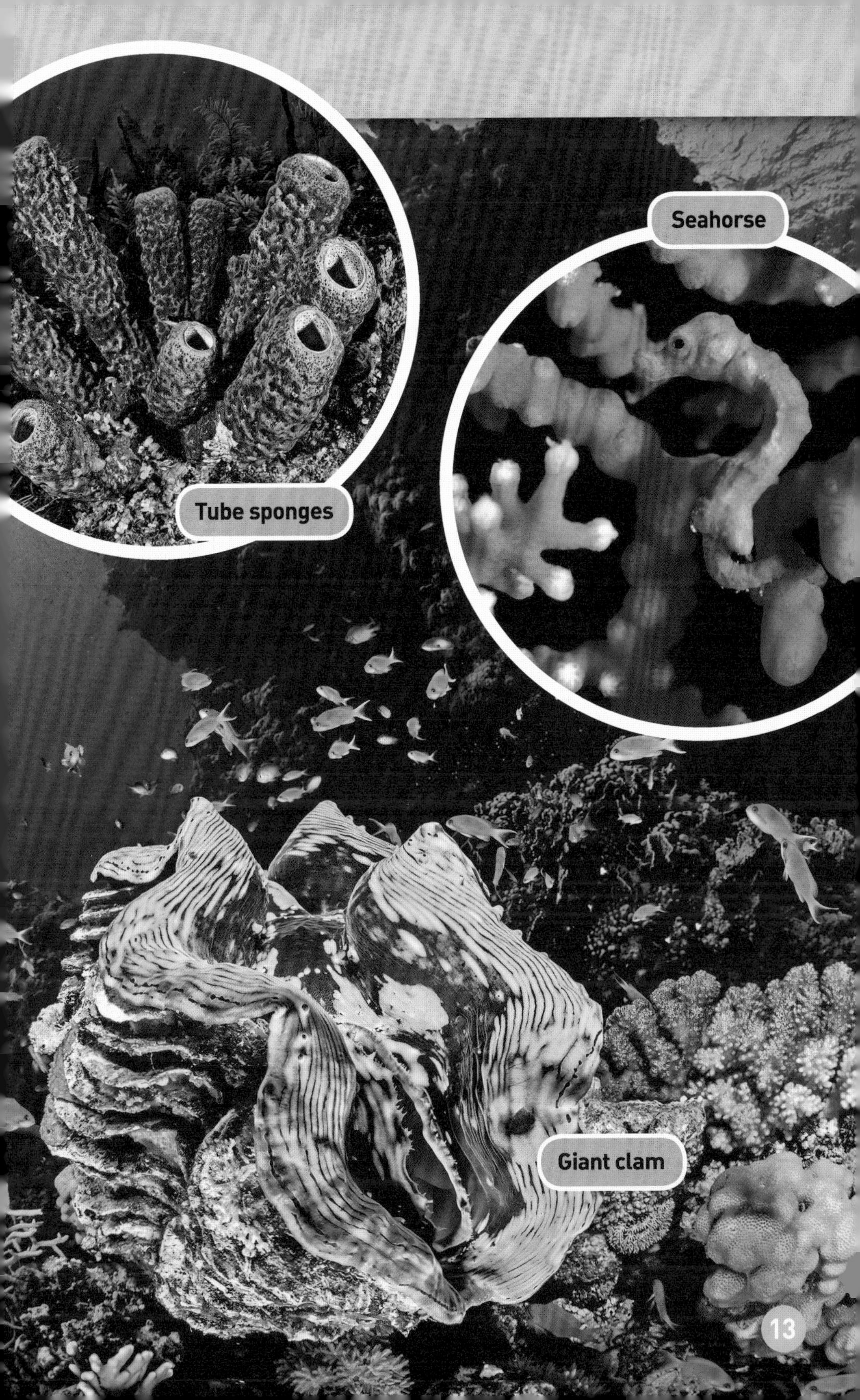
Seahorse
Tube sponges
Giant clam

Many animals use camouflage (KAM-uh-flazh) to hide along the reef. Some use it to stay safe from other animals that could eat them. Others use it to hide while they hunt.

Reef Talk

CAMOUFLAGE: An animal's natural color or shape that helps it hide from other animals

Reef stonefish

Trumpetfish

Cuttlefish

A stonefish's bumpy body blends in with the coral. A trumpetfish dives down and holds still. Its long, thin body stretches up like a tall sponge. A cuttlefish can change its shape and skin color to match the coral reef.

Reef Plants

Plants play an important role in coral reef ecosystems.

Tiny algae (AL-jee) live inside the coral polyps' soft bodies. The algae use sunlight to make food for the coral. This helps the coral grow.

Algae that grow inside coral polyps, as seen through a microscope

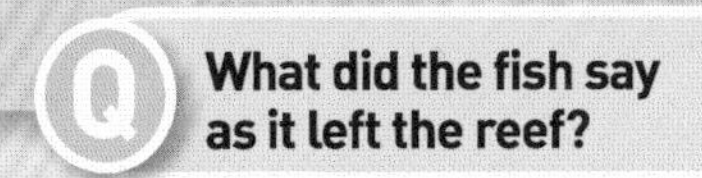

Sea grass helps trap mud from rivers that flow into the ocean. Sea grass also provides food for dugongs and sea turtles.

Reef Talk

ALGAE: A simple plant without stems or leaves that grow in or near water

The Reef in Darkness

As night falls, life along the reef changes. Different animals come out to find food.

Squirrelfish use their large eyes to search for shrimp in the darkness. Octopuses stretch their arms over the reef to feel for food. Sharks hunt for fish. Cone snails catch fish and worms.

6 COOL FACTS About Coral Reefs

1

There are more than 800 different kinds of hard coral in the world's oceans.

2

Brain corals can live for 900 years.

3

The first coral reefs on Earth formed about 600 million years ago, before dinosaurs were alive.

Most coral reefs today are between 5,000 and 10,000 years old.

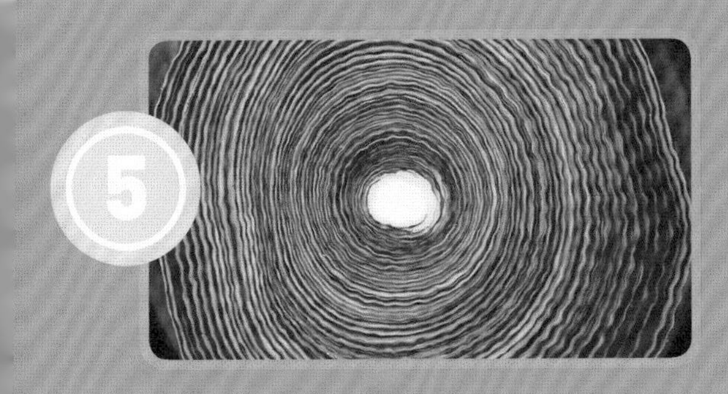

Corals have growth rings, just like trees.

The Great Barrier Reef is off Australia's northeast coast. It is the largest coral reef system on Earth. It can be seen from space!

People and Reefs

Coral reefs are important to more than just animals and plants. They are important to people, too.

Millions of people eat the fish that live along reefs.

Q Which coral is the smartest?

A A brain coral.

Many people earn money from fishing or taking tourists to visit reefs.

Reefs also help protect people and houses on land. They block big waves from crashing on the shore.

Rescuing Reefs

Many reefs are in danger. Scientists are worried about threats to coral reefs.

Ocean waters around the world are getting warmer. Coral polyps die when the water is too warm.

Pollution (puh-LOO-shun) sometimes spills into the oceans. It can harm reefs.

Fishing and boating can also damage fragile reefs.

Reef Talk

POLLUTION: Harmful matter that makes water, soil, or air dirty

But there is good news. Many people are working to save reefs.

Volunteers help clean up pollution on land and at sea. Some countries have special areas, called preserves, where coral reefs are protected.

Divers and swimmers—like you—can help, too! The next time you see a beautiful coral reef, look but don't touch.

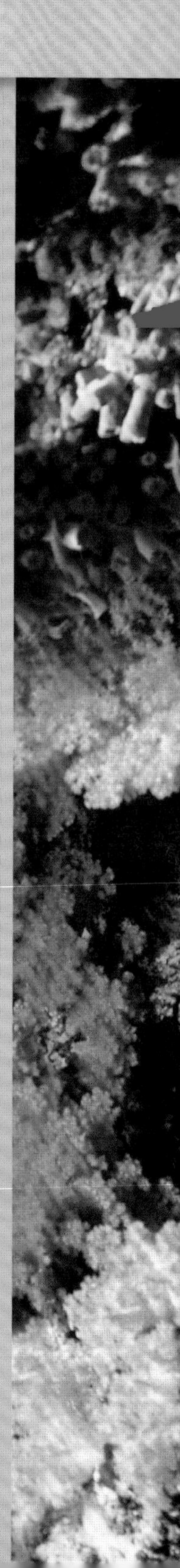

Q
Why did the diver stay away from the reef?
A
It seemed rather fishy.

QUIZ WHIZ

How much do you know about coral reefs? After reading this book, probably a lot! Take this quiz and find out.

Answers are at the bottom of page 29.

1

What are coral reefs made of?

A. Hard coral polyps
B. Rocks
C. Soft coral polyps
D. None of the above

2

Which of these animals does not live along a coral reef?

A. Eel
B. Horse
C. Sea star
D. Giant clam

3

Which of these animals can be found along reefs at night?

A. Cone snail
B. Octopus
C. Shark
D. All of the above

Earth's largest coral reef system is off the coast of which country?

A. Australia
B. Belize
C. Indonesia
D. United States

4

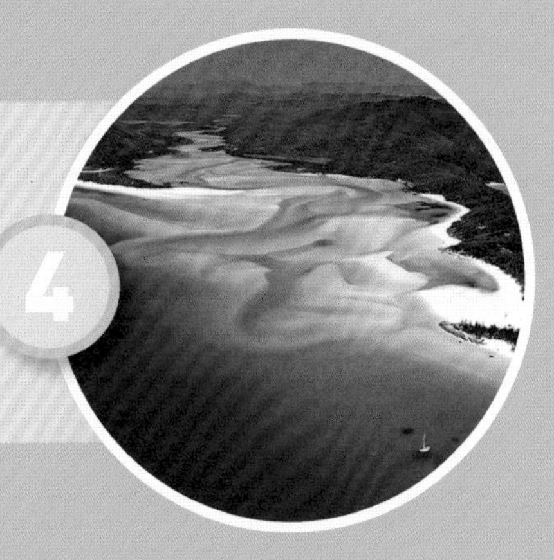

5

How old are most coral reefs today?

A. 5,000 to 10,000 years old
B. 50,000 years old
C. 500,000 years old
D. 500 million years old

What can be harmful to coral reefs?

A. Fishing and boating
B. Pollution
C. Rising water temperatures
D. All of the above

6

Answers: 1) A, 2) B, 3) D, 4) A, 5) A, 6) D

Glossary

ALGAE: A simple plant without stems or leaves that grow in or near water

CAMOUFLAGE: An animal's natural color or shape that helps it hide from other animals

CORAL POLYP: A small, simple sea animal with a tube-shaped body and a mouth ringed with tentacles at the top

ECOSYSTEM: All the living and nonliving things in an area

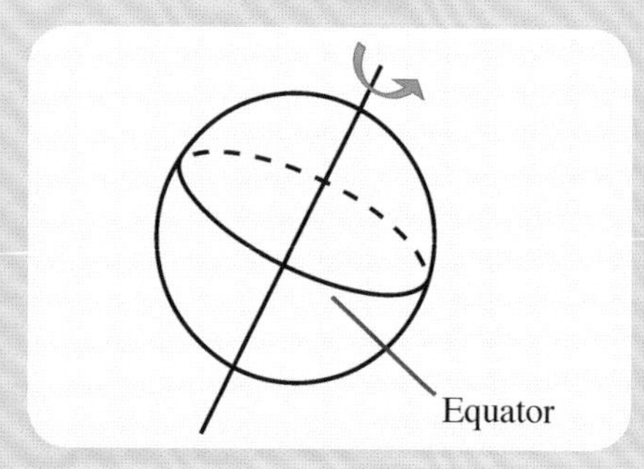

EQUATOR: An imaginary line around Earth halfway between the North and South Poles

POLLUTION: Harmful matter that makes water, soil, or air dirty

▶ 第 4—5 页

海底城市

浅海区看起来很平静。但是在水面下，珊瑚礁是一个忙碌的地方。

上百种甚至上千种不同的生物在珊瑚礁边上游动、藏身。它们在这里觅食、居住。它们在这座“海底城市”安家落户。

▶ 第 6—7 页

珊瑚礁是非常重要的生态系统。生活在珊瑚礁边上的海洋生物远比海洋里的其他地方多。

珊瑚礁遍布世界各地。大多数珊瑚礁生长在赤道两侧干净的浅海区。它们整年都要有阳光和温暖的海水才能存活。

珊瑚礁小词典

生态系统：一个区域所有的生物以及非生物

赤道：位于北极和南极正中间、环绕地球的一条假想线

▶ 第 8—9 页

珊瑚礁建造者

珊瑚礁看上去像是由岩石构成的。但事实上，它们是一群群叫“珊瑚”的动物。每个珊瑚群都是由许多独立的珊瑚虫堆积而成的。

在珊瑚礁生态系统中有两种珊瑚：石珊瑚和软珊瑚。只有石珊瑚虫才能堆积成珊瑚礁。它们因为柔软的身体上面长着坚硬的骨骼而得名。

一群珊瑚虫

珊瑚虫近图

珊瑚虫：一种小型的、简单的海生动物，身体呈管状，口部顶端周围长着触手

石珊瑚堆积成珊瑚礁。

软珊瑚不能堆积成珊瑚礁。

▶ 第 10—11 页

珊瑚虫可以和针尖一样小。但当许多珊瑚虫聚在一起时，它们可以堆积成绵延几英里（1 英里约等于 1.61 千米）的珊瑚礁。

新的珊瑚虫会在已有的珊瑚虫上长出坚硬的骨骼。许多年以后，一层层骨骼慢慢变成珊瑚礁。

口部张开的珊瑚虫堆积而成的扇柳珊瑚

石珊瑚礁

第 12—13 页

珊瑚礁上的邻居

在世界各地的珊瑚礁上可以看到大大小小的生物。海星沿着珊瑚礁的表面缓慢地移动。巨蚌也在那里休息。管状海绵像小烟囱一样从珊瑚礁上伸出来。海马把尾巴卷在珊瑚的碎片上。海龟在珊瑚礁边上游动。鳗鱼藏在珊瑚礁的缝隙里。

▶ 第 14—15 页

很多动物通过伪装隐藏在珊瑚礁边上。一些动物用它来保护自身的安全，以免被别的动物吃掉。另一些动物用它在捕猎时隐藏自己。

珊瑚礁小词典

伪装：动物天然的颜色或形状，帮助它不被别的动物发现

玫瑰青鲉凹凸不平的身体与珊瑚融为一体。管口鱼头向下潜入水中，静止不动。它细长的身体向上伸出去，就像高高的海绵。乌贼可以变换形状和肤色，以便与珊瑚礁相配。

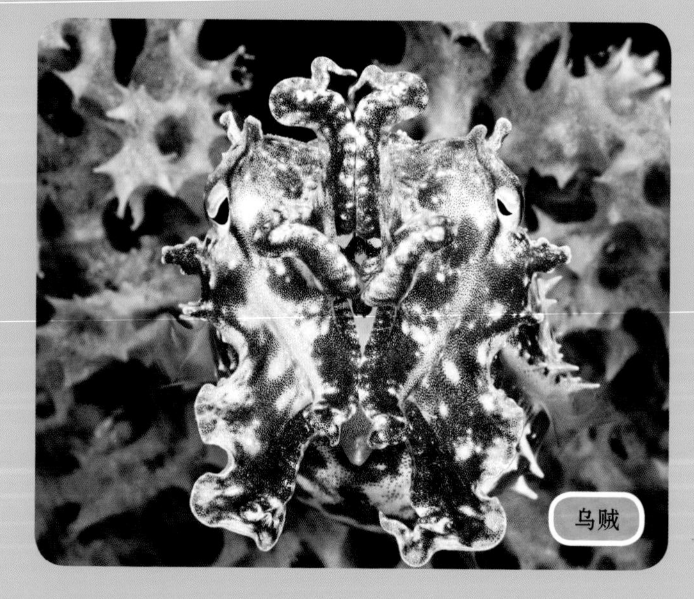

第 16—17 页

珊瑚礁上的植物

植物在珊瑚礁生态系统中起着重要作用。

微小的藻类植物生活在珊瑚虫柔软的身体里。藻类植物利用阳光为珊瑚生产食物。这有助于珊瑚成长。

海草有助于阻止河里的泥土流入海洋。海草也为儒艮和海龟提供食物。

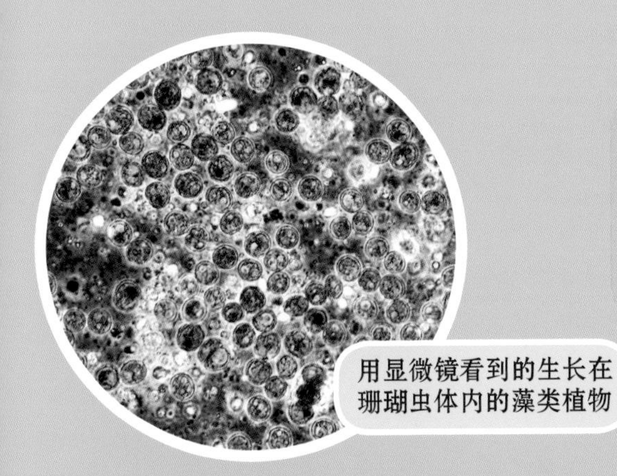
用显微镜看到的生长在珊瑚虫体内的藻类植物

珊瑚礁小词典

藻类植物：生长在水里或水边、没有茎或叶的简单植物

儒艮

▶ 第 18—19 页

黑暗中的珊瑚礁

夜幕降临，珊瑚礁边上的生活变了。各种动物都出来觅食。

金鳞鱼用它们的大眼睛在黑暗中寻找小虾。章鱼把腕足伸过珊瑚礁感知食物。鲨鱼捕鱼。鸡心螺捕捉鱼和蠕虫。

▶ 第 20—21 页

关于珊瑚礁的 6 件酷事

1 全世界的海洋中有超过800种石珊瑚。

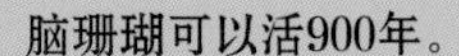

2 脑珊瑚可以活900年。

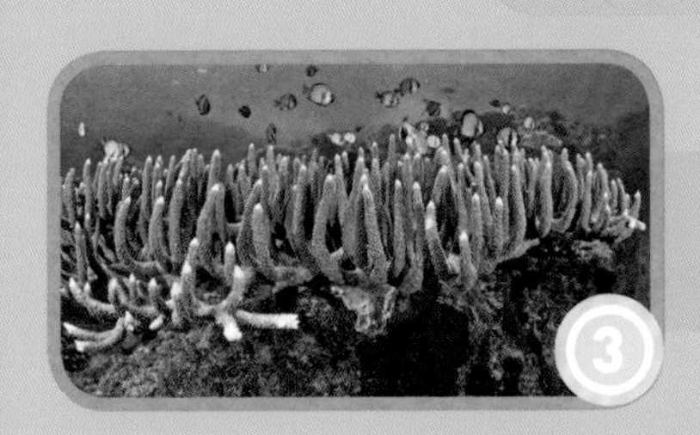

3 地球上最早的珊瑚礁形成于大约6亿年以前，在恐龙出现之前。

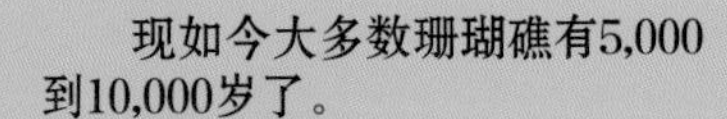

4 现如今大多数珊瑚礁有5,000到10,000岁了。

5 和树一样，珊瑚也有年轮。

6 大堡礁位于澳大利亚东北海岸附近。它是地球上最大的珊瑚礁生态系统。从太空中可以看到它！

第 22—23 页

人与珊瑚礁

珊瑚礁不仅对动物和植物来说非常重要，它们对人类来说也很重要。

很多人吃生活在珊瑚礁边上的鱼。

许多人靠捕鱼或为参观珊瑚礁的游客做向导谋生。

珊瑚礁还有助于保护陆地上的人和房子。它们阻拦冲向海岸的巨浪。

第 24—25 页

拯救珊瑚礁

死亡的珊瑚

许多珊瑚礁都处于危险之中。科学家为珊瑚礁受到的威胁感到忧虑。

全世界的海域正在变暖。水温过高，珊瑚虫便会死亡。

有时候污染物会流进海洋。它会危害珊瑚礁。

捕鱼、划船也会伤害脆弱的珊瑚礁。

珊瑚礁小词典

污染物：弄脏水、土壤或空气的有害物质

第 26—27 页

但还是有好消息。很多人都在致力于拯救珊瑚礁。

志愿者们帮忙清理陆地上和海洋里的污染物。有些国家还设立了名为“保护区”的特殊区域，那里的珊瑚礁得到保护。

潜水者和游泳者——比如你——也可以帮忙！下次当你看到一片美丽的珊瑚礁时，看看就好，千万不要摸它。

▶ 第 28—29 页

答题小能手

关于珊瑚礁你了解多少呢？看完这本书之后，可能很多吧！测试一下就知道了。答案在第 29 页下方。

1. 珊瑚礁是由什么构成的？

A. 石珊瑚虫　　B. 岩石
C. 软珊瑚虫　　D. 以上都不是

2. 下列哪种动物不生活在珊瑚礁边上？

A. 鳗鱼　　B. 马
C. 海星　　D. 巨蚌

3. 夜里在珊瑚礁边上可以看到下列哪种动物？

A. 鸡心螺　　B. 章鱼
C. 鲨鱼　　D. 以上都是

4. 地球上最大的珊瑚礁生态系统在哪个国家的海岸附近？

A. 澳大利亚　　B. 伯利兹
C. 印度尼西亚　　D. 美国

5. 现如今大多数珊瑚礁已经多少岁了？

A. 5,000 到 10,000 岁　　B. 50,000 岁
C. 500,000 岁　　D. 5 亿岁

6. 哪项可能会危害珊瑚礁？

A. 捕鱼和划船　　B. 污染物
C. 水温上升　　D. 以上都是

第 30 页

词汇表

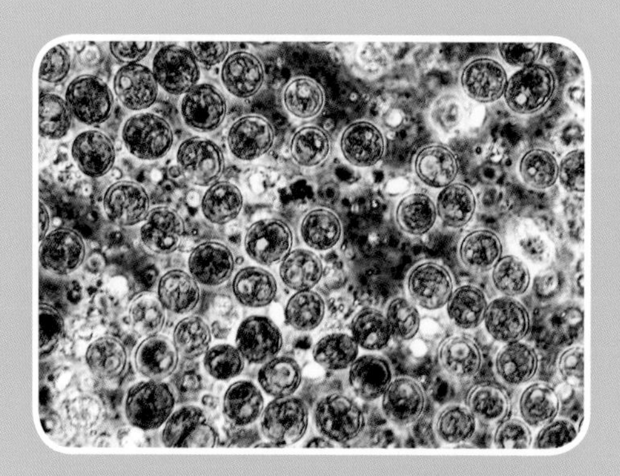

藻类植物：生长在水里或水边、没有茎或叶的简单植物

伪装：动物天然的颜色或形状，帮助它不被别的动物发现

珊瑚虫：一种小型的、简单的海生动物，身体呈管状，口部顶端周围长着触手

生态系统：一个区域所有的生物以及非生物

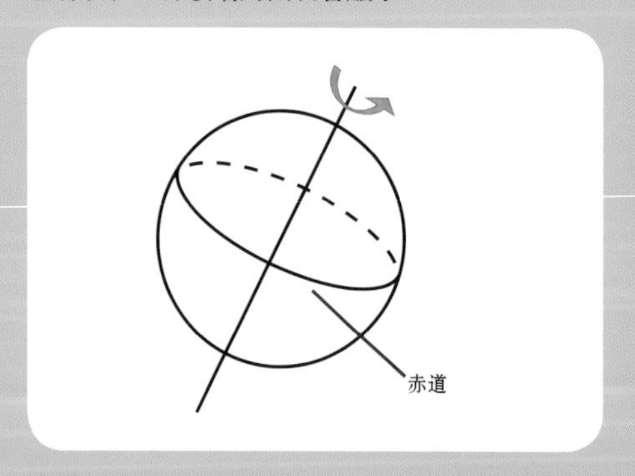

赤道：位于北极和南极正中间、环绕地球的一条假想线

污染物：弄脏水、土壤或空气的有害物质